Marc Goldstein

LIVING AND MEASURING

Contents

Penguin Books

We measure with our bodies. A long time ago, men measured distances by arm, hand, foot lengths (feet) and paces: children still do this for various games. Even when we measure with instruments such as rulers, scales and even computers, it is still a part of our body – the eye – that *sees* the measurement and another part – the hand – that *writes* down the answer.

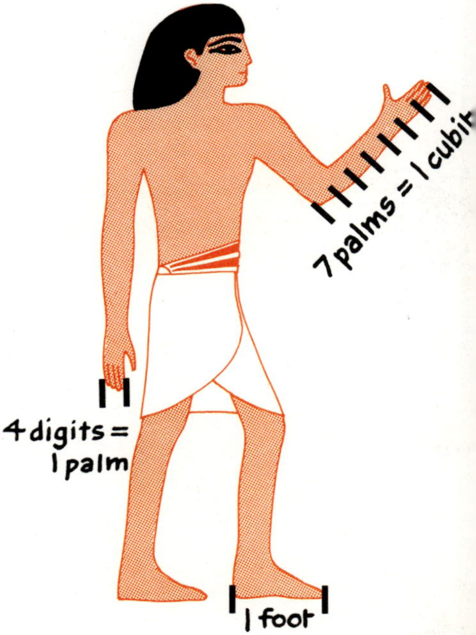

And we measure *for* our bodies (for example when we work out the size of clothing or the amount of food for a meal).

In all we do we are constantly comparing amounts, noticing how things arrange themselves and making things of the right size. In this book you will read how much of this way of treating things, which we call *mathematics*, is connected with our bodies.

You in your home

'It's good to be home – I'm whacked!'

'Oh my poor feet!' You often hear adults saying this, and children too, sometimes. People who have been standing or walking for a long time get tired and their feet ache. This is not surprising when you think of the great weight that is pressing down on our feet. The head, the trunk, the upper limbs (the arms) and the lower limbs (the legs) – all these parts weigh down on our feet! Which do you think is the heaviest part?

To answer this question you would need to know the weight of each separate part of the body. You could ask someone or look it up in a book, but some of you may wish to do the weighing yourselves. You could think of several ways of doing this with different kinds of scales.

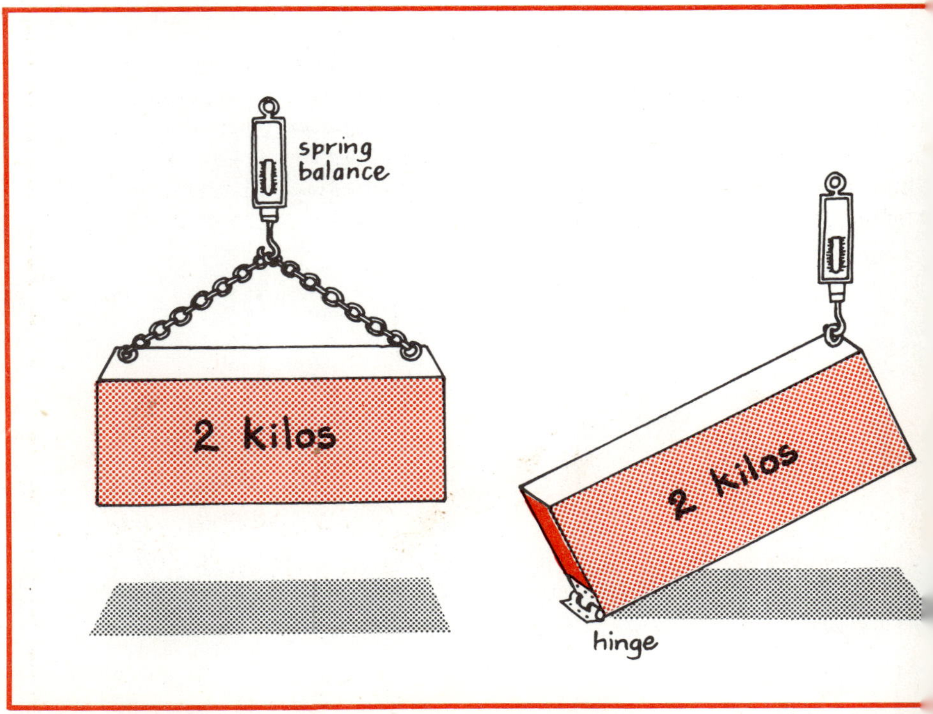

Here is one you might like to try. It is not absolutely exact but it will give you a fair idea of how your body weight is shared out. It uses a *spring balance* and weighs by *lifting.*

I don't expect any of you had thought of cutting off one of your arms or legs! So whatever section you weigh will remain hinged to the rest. Therefore there are two things to remember. One is that the person being weighed must not use his muscles to resist or to help but should lie quite relaxed. The other thing is that when you are weighing something which is hinged down at one end, the *point* at which you lift is very important. The illustrations and experiment below will help to explain this.

ll you need is a length of wood – it does not have to weigh 2 kilos – can be any weight. (If it *is* too heavy for one spring balance you ould use *two.* Can you think how this is done?) Only two of these pring balances will show 2 kilos, the actual weight of the wood. /hich are they? When weighing an object that is hinged at one end, t what point along the object should we fix the spring balance?

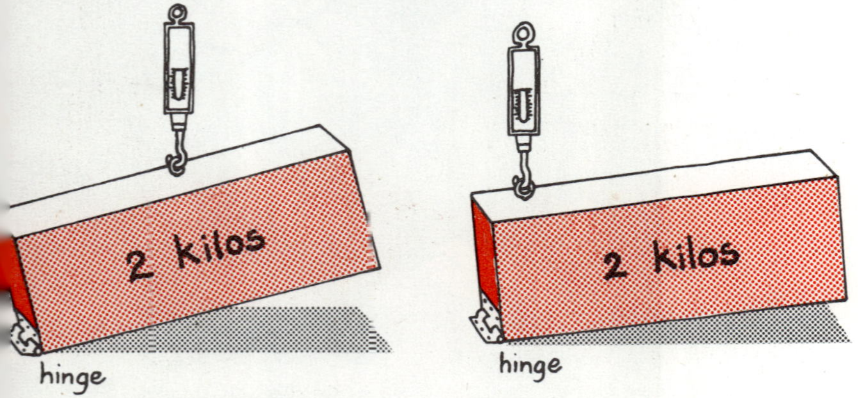

Weighing a person in parts

These pictures show a girl weighing her friend in parts. Think of the experiment with the hinged piece of wood and see if she has fixed the scarf in the right places.

When the head is weighed the person lies face-upwards.

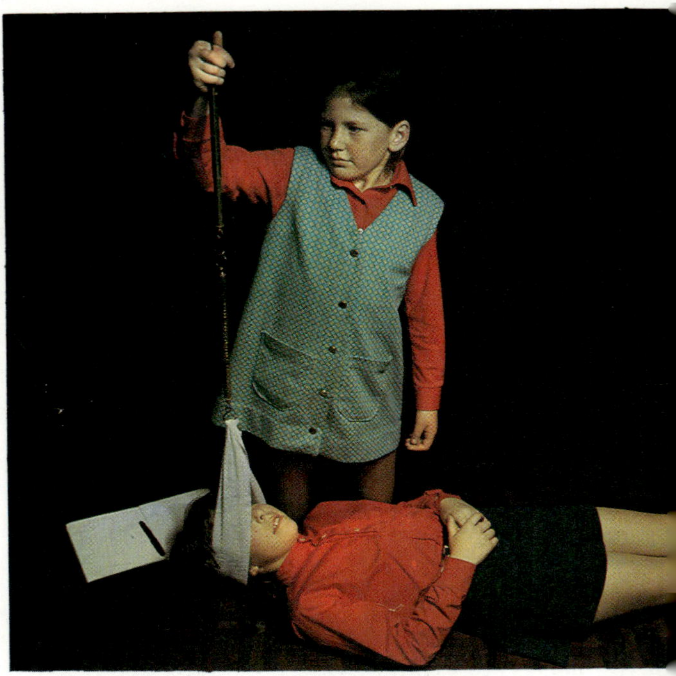

The arms are weighed in the same face-upwards position.

For weighing the legs the person lies face-downwards. Do you know why she has changed her position?

The girl has found the weight of her friend's head, arms and legs. Now she is finding the weight of her trunk by taking away the weight of her head, arms and legs from the weight of *all* her body.

Are you sitting comfortably?

'Take the weight off your feet!' This is a friendly way of saying 'sit down'. But it happens to be quite true. As soon as you sit down, a great part of your weight is taken off your feet. The total weight of your trunk, arms and head, and the upper parts of your legs (thighs) are no longer pressing down on your feet.

Where are they pressing? If you are trying this out as you go along, you will be sitting firmly on your seat by now! But is there now no pressure on your feet? This depends on the kind of chair you are sitting on. Look at these pictures.

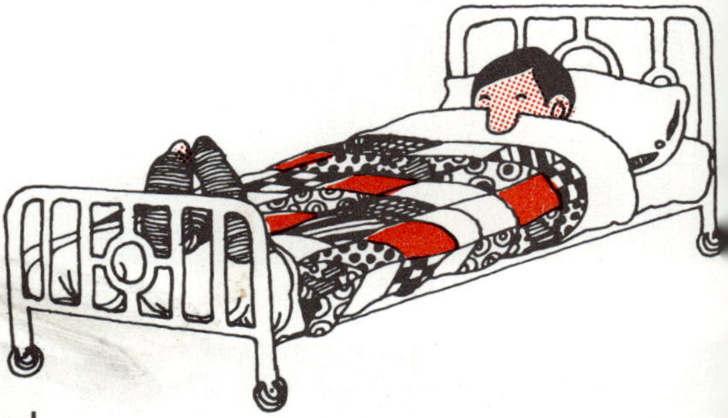

In bed:
The whole weight is 'shared out' along the body from head to heels.

On a high chair:
The head, trunk and arm weight are on the seat, the leg weight is on the feet. By using bathroom scales, can you find out how much weight the feet are bearing now?

Lower chair:
How much weight
is on the feet?

Easy chair:
And now, how much
weight is still on the feet?

Which is most comfortable and which is most uncomfortable?

If you are sitting at a table and need to lean forward to eat, write or draw, which would be the best chair to use?

Thinking about particularly high chairs, why would you think a kitchen stool is higher than a kitchen chair? And at refreshment bars; why are the chairs so high there? Have you noticed where the customer rests his feet?

Just an average chair

The designing of chairs for different purposes is very interesting and new designs are being thought up all the time. Have you ever thought how comfortable the tiny chair on a bicycle is? Of course, it is called a saddle but, although it is very small compared with any other kind of chair, it is extremely comfortable. The man who invented it thought up a wonderful idea of getting exactly the right shape – he sat himself down on a heap of wet sand! When he got up, he knew that the dent he had left in the sand was the shape he needed to fit snugly on a bicycle.

But supposing the inventor was a large, broad man. Is it likely that the saddle he designed would be suitable for a small, thin man? For a perfect fitting saddle, each person who buys a bicycle would need to go and sit on a pile of wet sand and have his very own saddle made for him! This would be very expensive (as well as uncomfortable), so what we have on bicycles is an *average* saddle that may not be exactly right for everybody but is near enough to be quite comfortable.

A saddle for a pedaller

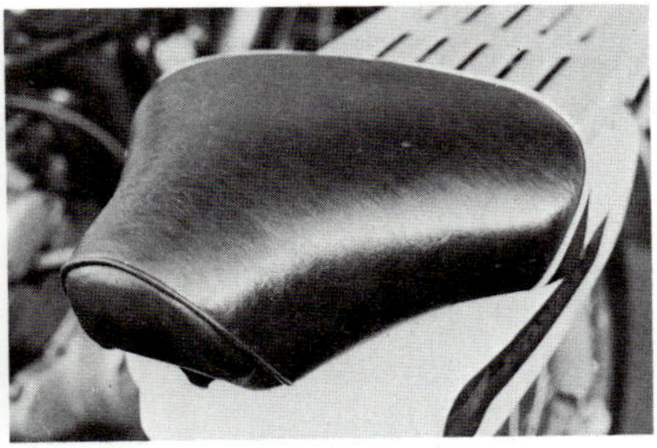

A saddle for a sitter

The same thing applies to ordinary chairs. You have already seen that the height of the seat is very important. So chair seats are made of an *average* height to suit most people who will sit on them. Look at this picture of a person sitting on an ordinary chair. It is just the right size because all his weight, except for his lower legs, is resting on the seat.

It is too much to expect that *everyone* would fit the chairs exactly, just like the saddle – they would need to be all the same sized people. You might like to find out whether the chairs in your classroom have a seat height that is the same, or higher, or lower, than the average seat height of the people who use them. How do you find the 'seat height' of a person? You get him to bend one knee so that he makes a 90° angle (which we call *right angle*) at the knee-joint and you measure the distance from the lower thigh to the bottom of the heel. Th s is clearly shown in the picture.

And the average person

Now if you note down the seat heights of all the children in your class, you can *add* the heights together to find the *total.* This great total height can be shared out fairly by dividing it equally among all the people who were measured and the result will be the average seat height of those people. You may need help from your teacher to do some of the working out but do check that the average you get is somewhere between the highest and the lowest seat heights of the people you measured. Now you can *compare* the average seat height of the children in your class with the height of the chairs they use. This is properly known as a 'mathematical experiment'. This title means that someone (in this case, you) has asked a question and tried to find out the answer for himself.

What was *your* question? It was: Do most of the children in my class sit lower or higher than their natural seat height? If you want to 'complete the experiment', write an account of what you discovered and how you found out, and illustrate it with diagrams or pictures. You can make a set of model chairs out of balsa wood or card, and small figures to sit in them to show the different *postures* (or body positions) for resting, working or eating and the type of chair that suits each one. You may even be able to make a model *adjusting* chair like a dentist's or a barber's chair.

Here is a simple model of a person that can be sat in your model chairs. He is made of cardboard and jointed with paper fasteners. If you make him to *scale,* a good one to use is 1 : 10. That means every length on your model will be one tenth of the length of the same part on a real person. So if the line from the top of a real man's head to his shoulder is 28 centimetres, the same line of your model's head will measure 2·8 centimetres. Of course, he will only fit your model chairs if they are made to the same scale.

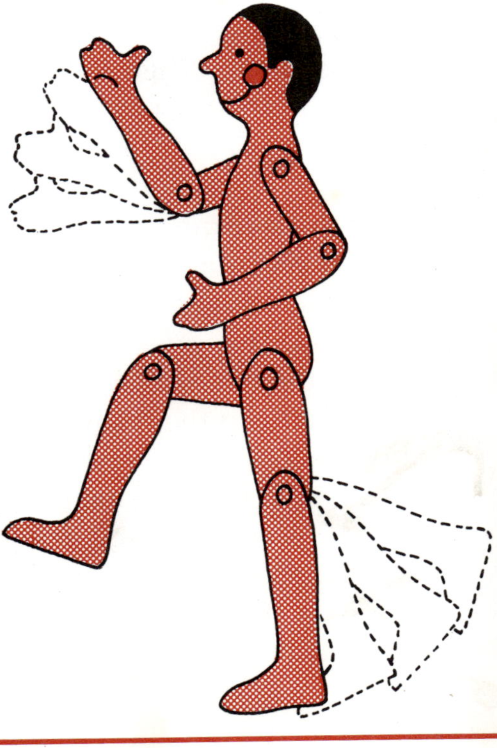

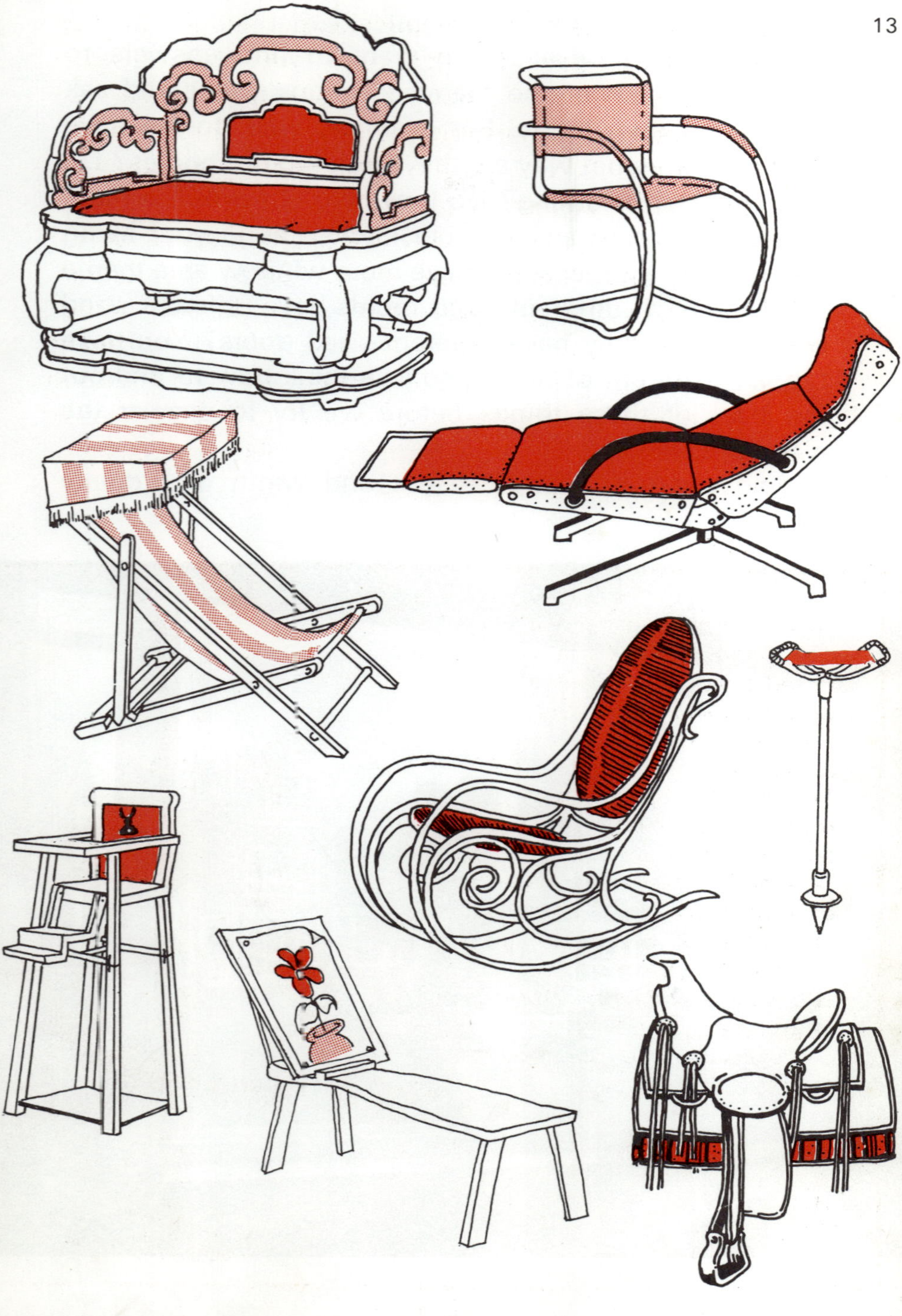

Doors, floors and windows

We experiment because we want to find answers to questions. Think of the many questions you could ask about your own home. For example, why do we have windows? In what way are they useful to our bodies? Do they help our breathing? Do they help our eyes? Is there as much window area in your kitchen as in your living room? Does a larger room have more window area than a small room? If different sized rooms have different sized windows, do they have different sized doors? You have probably run out of breath, so let's think how to find out the sizes of these things before we try to answer the questions.

Find out the average height and width of ordinary

house doors and ask yourself why they are that particular size. It seems natural enough to think that ordinary doorways are made the size they are so that normal people will be able to walk through them comfortably, without knocking their heads or becoming stuck in the opening. We read in history books that King Henry VIII became so fat that he could scarcely squeeze through an ordinary door, but maybe doors were smaller in those days. Have a look when you go to visit a Tudor house.

In some buildings, however, you will find doorways that are much higher than they need to be. This is because the designer or architect of the building wanted them to look beautiful, and if the whole thing was large and grand, he wanted the entrance to be grand too. Such doorways

Land window

Heavenly window

Water window

have nothing to do with the size of the people who go through but are meant to *look* right, to fit in with the rest of the building.

It is unlikely that you live in a large building such as a castle or a palace. Some people do and they often find things difficult for them. It is because they are surrounded by large *areas.*

In the picture, you can see how to find out the area of a room in an ordinary house. Once you know how to measure the area of floors you can measure the doors and windows we were talking about, or anything else with a flat surface. So now you can start to think about the questions on page 14.

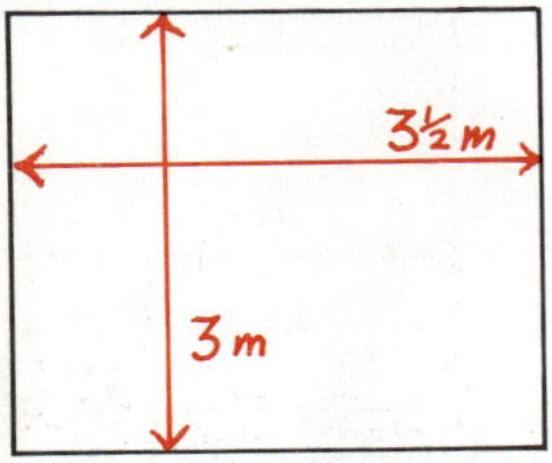

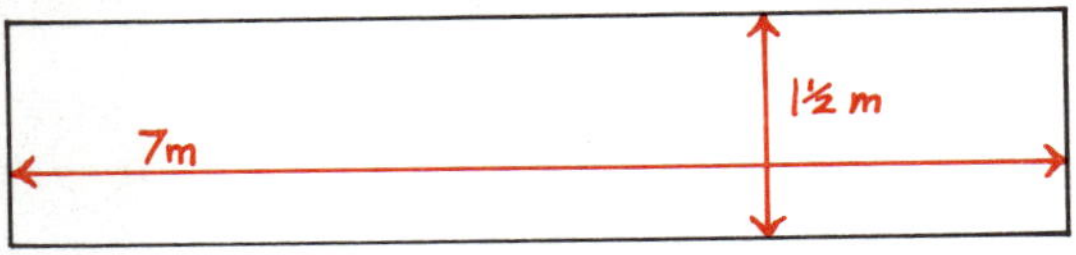

The *area* of a room is the amount of space it covers on the ground, that is its floor space. Here are the floors of the hall and one of the bedrooms of a house. Which area do you think is bigger?

Now we will 'square them off' into units of square metres.

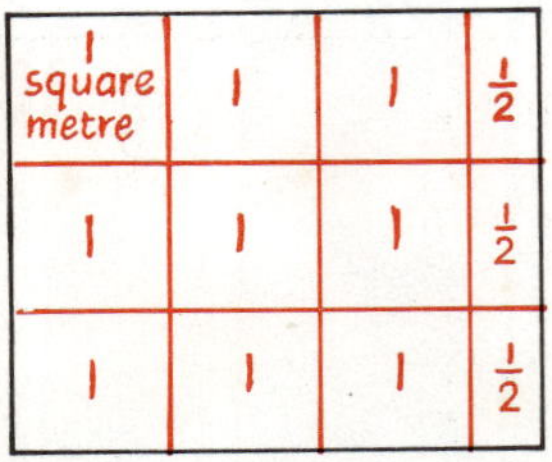

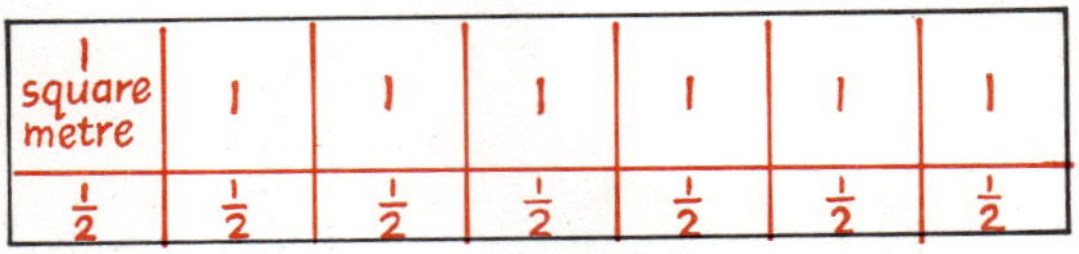

As you can see they have the same area.

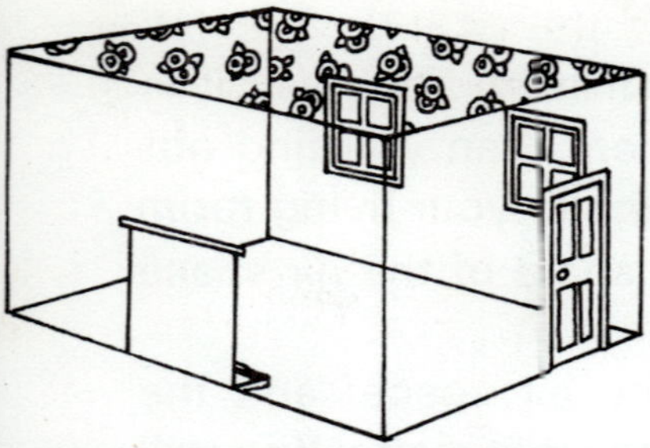

scale: 1 centimetre = 1 metre

Look at the drawing of this room. Below, it is shown so that you can see the whole area of the floor and walls laid out flat. The squares are to help you to compare the different areas. The side of each square is 1 centimetre long but it *represents* 1 metre on the actual room. What is the actual size of the floor and the height of the room?

You can find out the price of carpet, wallpaper and curtains. See how much it would cost for a room like the one here or for a complete house, or for a 'stately home'.

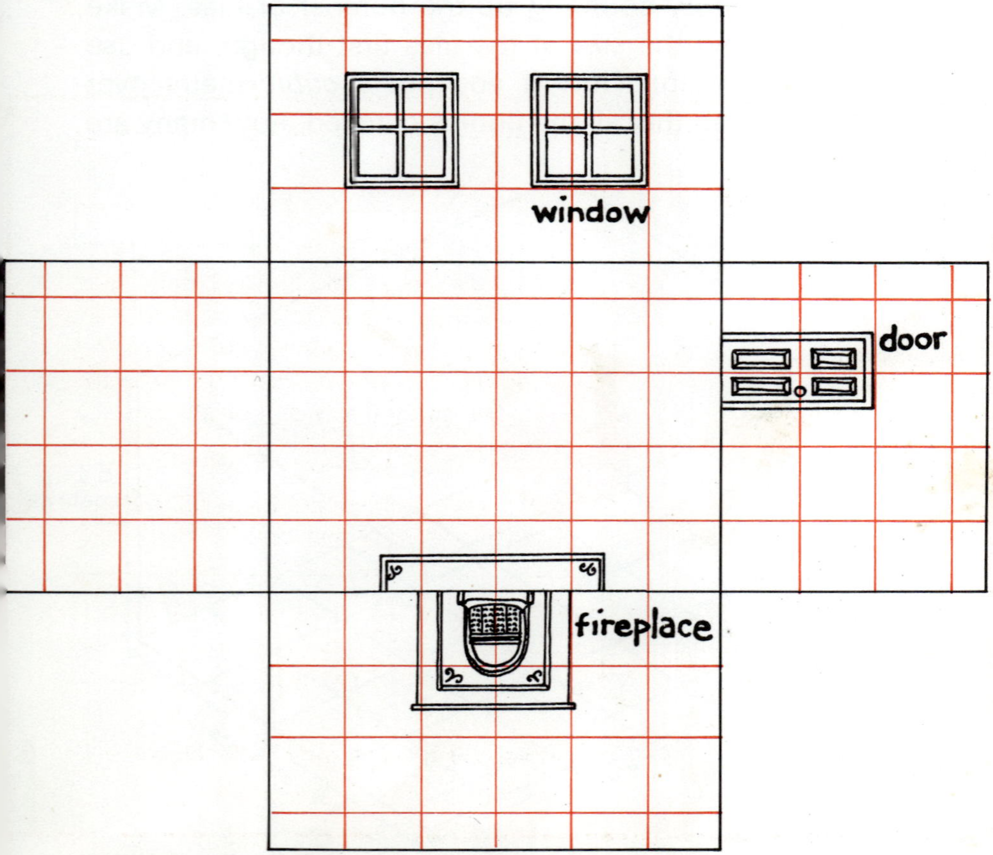

'And they all live in little boxes'

Many people prefer 'big boxes' to live in. They say it is pleasanter to be in a large, 'airy' room. Can you find out the size of *your* share of the air space in your living room at home and compare it with your share of the air space in your classroom?

To do this we must *measure* the air space; and the measuring unit we use for this is the *cubic metre.* You will see one in the picture, tucked right into the far corner of the room. A cubic metre is one metre long, one metre wide, one metre high.

You can divide the floor of this room into square metres as you did with the rectangles before. If you are lucky enough to have a tiled floor at home you can try this there just by counting up the number of tiles. Make sure you know the size of the tiles first though, and use that measurement. Now if you put a *cubic* metre over each square, so that all the floor is covered, how many are

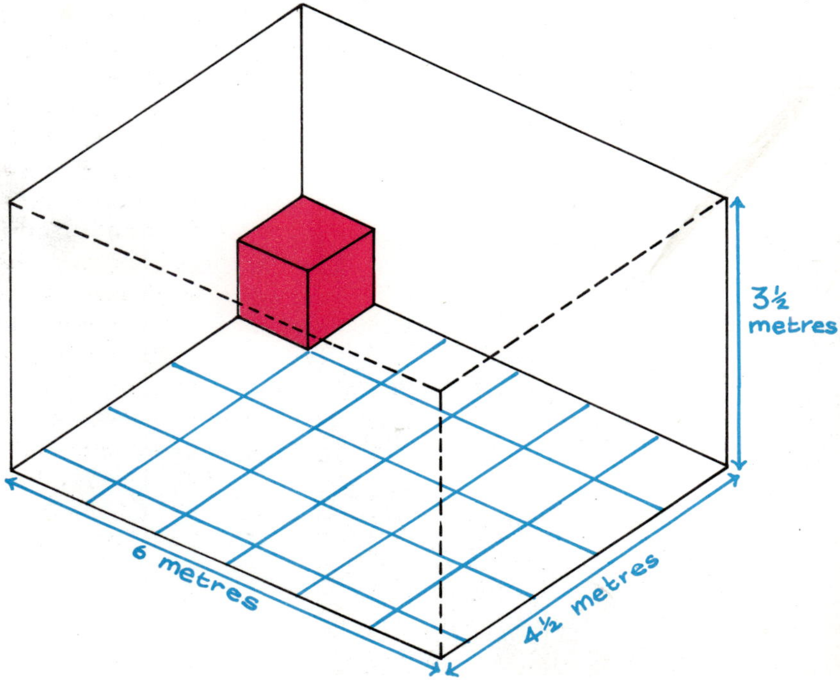

there? How far up the wall will this layer of cubic metres reach and how many cubic metres would fill the entire room?

If four people are in the room, they share the volume of air space among four.

How many cubic metres of air space for each of them?

With this method (or a better one if you can think of it) you should be able to find out the air space for each person in *any* room.

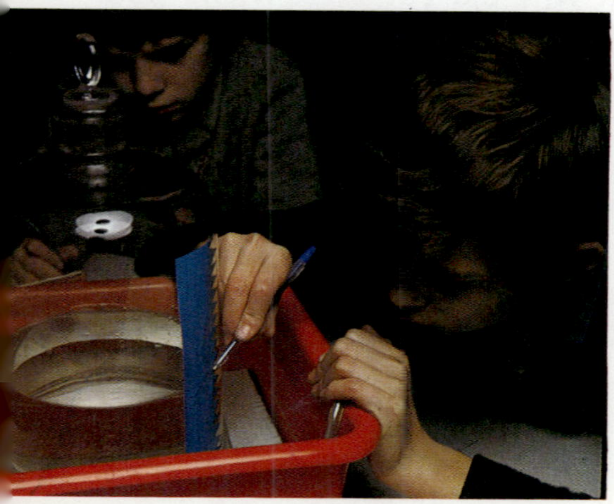

Using a measuring jar to find the volume of one breath

You can find a rough measure of a lungful of air by using a measuring jar and a length of tubing. The boys in the picture are using a bell jar, which is a jar with no bottom to it. A few layers of paper in the bottom of the bowl keep most of the water from escaping when the jar is filled. After replacing the stopper, one end of the tubing is slipped under the bottom rim of the jar. The water-level is marked off on a scale and then a normal breath is breathed into the tube. The new water-level is measured and written down. The volume of one breath can then be worked out. Find out how many times you breathe in one minute and work out the volume of air you breathe in this time. Compare this with the volume of air in the room.

'A bit chilly'

Have you ever thought that, in a way, your body is always making a hole in the air? You are surrounded by air, which is pressing against the whole surface area of your body. Although some parts of you are more *exposed* than others, ordinary clothing does not prevent air reaching your skin. What it does is to *hold in* a layer of air that has become warmed by your body. If the air in a room is really cold, it slowly passes through your clothes, then it cools down this layer of warm air and you begin to feel chilled. So when we say, 'This room feels cold', what we really mean is that the *air* in the room feels cold on the surface area of our bodies. In winter, once the air in a room has been warmed, we try to keep it in by shutting doors and windows. But the warm air can become cold all the same. Where the air in a room is resting against a cool area, such as a cold wall or window, the air will begin to cool down, like the air next to your body did. When you are in a warm room, try the experiment of resting the back of your hand against an inside wall, then against an outside wall (this does not mean going outside) and then against a window pane. Which of those surface areas will be cooling down the air in the room most quickly? I think you will find it is the window pane. This would explain why old cottages, where poor people lived in the past, often had tiny windows. They had to choose between small windows and small fires in winter or big windows and large, expensive fires.

In those days, when cottage windows were so small, churches had vast windows. This made them very cold in winter but, of course, men did not design churches for living in. They cared more about letting in plenty of daylight and making the building beautiful than with keeping the place warm.

Today most people seem to prefer large windows. If they have them, they must make sure that they do not cool down the warm air inside the room in winter. Thick curtains help to keep in the warmth but, of course, they shut out the daylight. Some houses have *double glazing* which means that their windows consist of *two* panes of glass with a small air space between. This air space acts as a sort of cushion between the warmth inside and the cold outside. And daylight comes into the room just the same. So the area of windows is important. A room with a lot of window area will need a lot of window *insulation* (curtains or double glazing). But even if the window insulation is poor, a room can be kept warm if it is heated enough (of course, if you have to keep using a lot of heat to make up for what is being lost through bad insulation, it will be expensive).

A double-glazed entrance keeps out the cold.
Would it be useful in hot lands?

Nowadays *central heating* has become very popular. Water is heated up in a boiler, and it then flows along pipes to *radiators* in various parts of the house.

The radiators are designed to spread out to as big an area as possible because the larger the hot area, the more warmth will pass into the air in the room. People whose job it is to install central heating have worked out what surface area of radiator is needed for each size of room. It does not matter whether the surface area is made up of one large or several small radiators as long as the *total surface area* is sufficient. Can you think of a way to measure the surface area of the radiators in your school, if you have them?

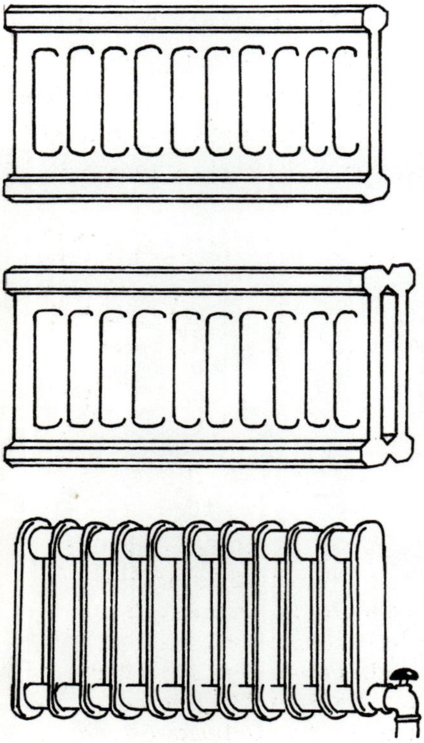

1 Single-panel radiator
The surface area is the total area all round, *front and back.*

2 Double radiator
This will have twice the surface area of the radiator above.

3 Hospital radiator
To find the surface area of this type is difficult but it can be done if you deal with it in sections. Measure one section and draw a plan of it as it would look if you cut it open and laid it out flat. Square it off, as you did with the house on page 16, and find its area in square metres. Then multiply your answer by the number of such sections that make up the whole radiator. (You will probably find that the hospital radiator has a greater surface area for its size than a single-panel radiator of the same length. Why is this?)

If you flattened out this piece of paper you would find that it would be exactly the same size as a *page* from this book. It has the same *surface area*. Now imagine all the pages in this book crumpled up separately and put into a box. (You could cut up some newspaper to the same size and actually do this.) Compare the total surface area of these crumpled pages with the surface area of the book when it is closed with all the pages flat. Which pages will have the greater amount of air surrounding them – those in the closed book or the crumpled ones in the box?

The crumpled-up paper might remind you of a sponge, like the lung below does, too. Would you expect there to be a great deal of surface area inside a sponge?

'All crumpled up'

Here are some parts of the inside of you that look 'crumpled', but there are good reasons for this. For example, the lungs. When we breathe in, our lungs fill with air. The millions of tiny parts of oxygen in that air have to pass into the blood through a thin, skin-like wall.

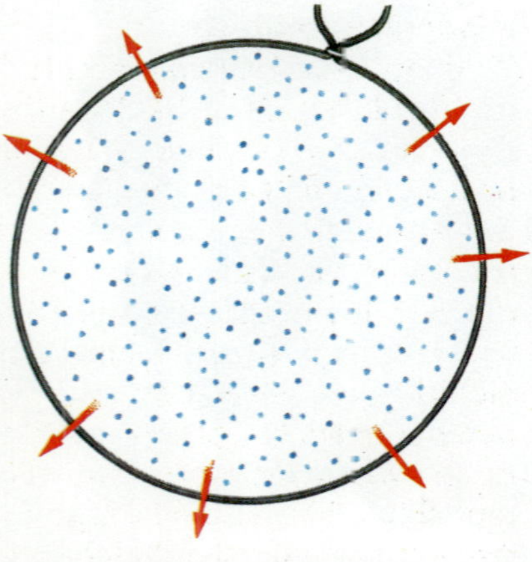

If our lungs were simply like large hollow balloons, it would take far too long for the atoms of oxygen in the middle to get to the wall.

This piece of lung looks much more like a sponge than a balloon. It branches out into thousands of tiny 'balloons' with a great area of wall all bunched up. The parts of oxygen do not have to wait but can pass through quickly into the blood.

24 Here are two brains. One is human, the other is the brain of a parrot. Notice how much more convoluted (crumpled or crinkled up) one is than the other. The human brain, has a much greater 'thinking area'.

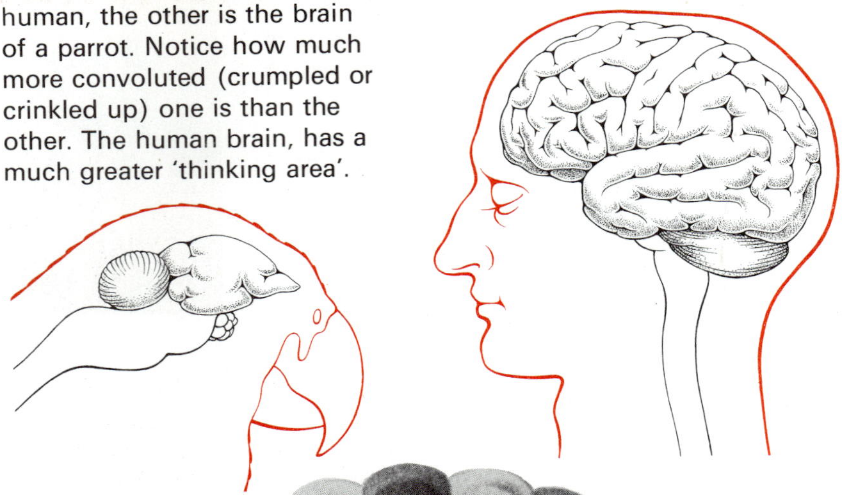

Parachutes are packed into spaces not much bigger than the area which holds our brain.

You and your clothes

In the last section we mentioned that the clothes we wear hold in to our bodies a layer of warm air. This might give you the idea that it is the purpose of all clothing to keep us warm. But many parts of the world are hot lands and there clothing is often used to *shield* the body from the heat of the sun, for, if the air near the body is cooler than the air outside, the clothes will stop the *cool* air from escaping. Of course clothes should also *decorate* the wearer – we talk about a person being 'well dressed' or 'beautifully dressed'.

There are two important things about clothes: what

they are made of and how they are shaped. Prehistoric man would kill an animal, cut off its furry skin and tie it round his body for clothing. But, because the animal had to be cut open in order to get the pelt off, and in those very early days no one understood how to sew, the result was a very rough and ready garment with great gaps that would let in the cold. Later on men discovered how to make needles out of bone and use animal sinews as thread, and so the art of tailoring began. This art consists of cutting a shape out of a sheet of material in such a way that, when it is stitched together, it fits round the body.

A perfect fit

Now if we were all cube shaped, this would be quite easy. The tailor would just have to measure the length of one of our edges and then draw a shape like this:

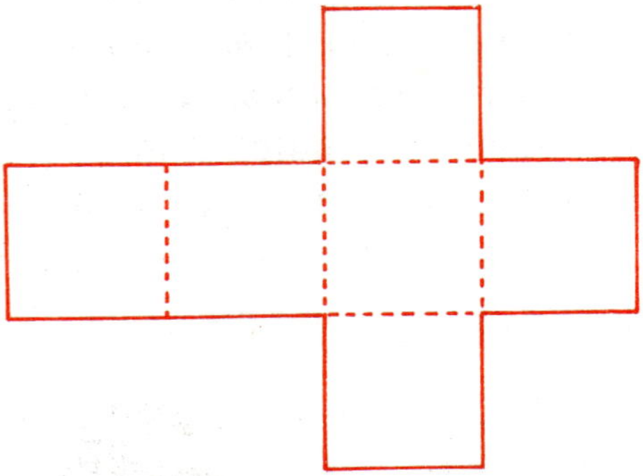

This is called the 'net' of a cube. When the edges were sewn together and holes cut out for head, arms and legs, this is how Mr Cube would look.

(He would probably find it difficult to get inside his new coat though!)

But your body is nothing like a cube. If you think about it, you will probably agree that it is much more like a cylinder. So are your arms and legs. The 'net' of an open cylinder is very easy: rolled out flat, it is simply a rectangle.

When you join edge A to edge B it looks like this:

Now if you make thinner cylinders for arms, cut two armholes and stitch the parts together, you are beginning to get something like a coat shape.

Can you see two things that are wrong here? One is that there is no covering for the shoulders. The other is not so obvious. It has to do with the sleeves. If a person put his arms through these sleeves and then let his arm drop, a great fold of the sleeve material would wedge itself into his armpit. Can you explain, with the help of this drawing, how this can be avoided?

'Setting in' a sleeve, as tailors call it, is a very skilful job, but there are many other problems to solve if a garment is to fit closely to the body, because our shape is not simple but is made up of many different curves. You may be interested to find out the effects you get from sewing together various straight and curved seams. Here are just a few suggestions but you will probably think up more of your own.

'Sausage' shape
(straight seam)
Sew edge A to edge B, stuff with waste material and tie the ends tightly.

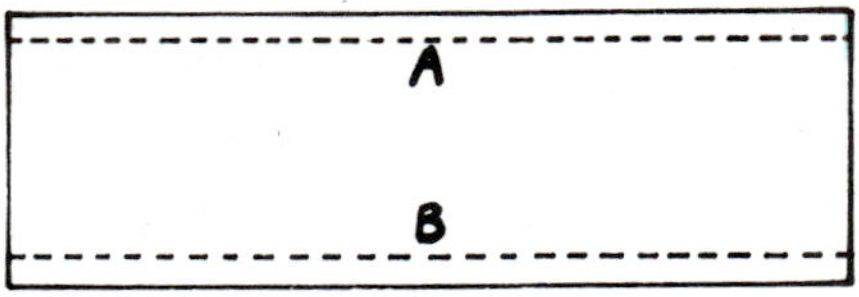

Result

'Skull cap' (curved seam)

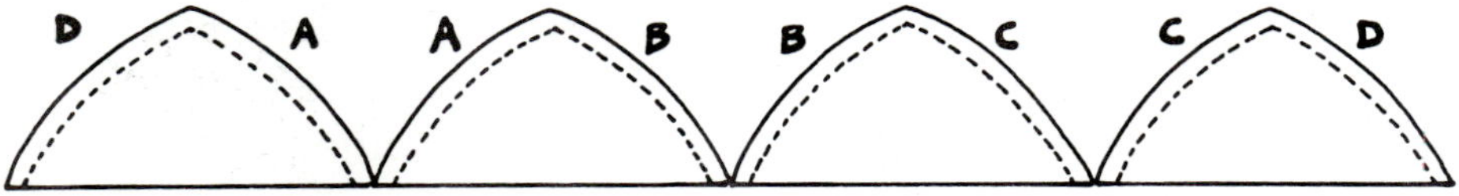

Sew A to A, B to B, C to C, D to D.

Result

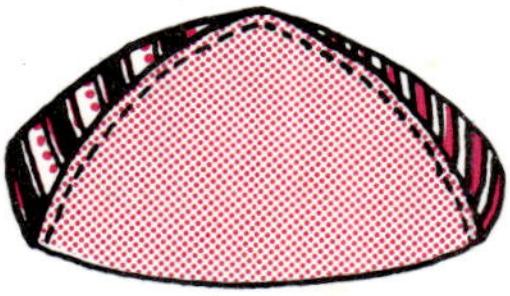

'Pyramid' (straight seams)

You can see where to stitch. But can you also see how, if your piece of material is big enough, you can save yourself work by drawing the *total* shape, so that you will have only four seams to sew? Stuff the shape with waste material before finishing the last seam.

Result

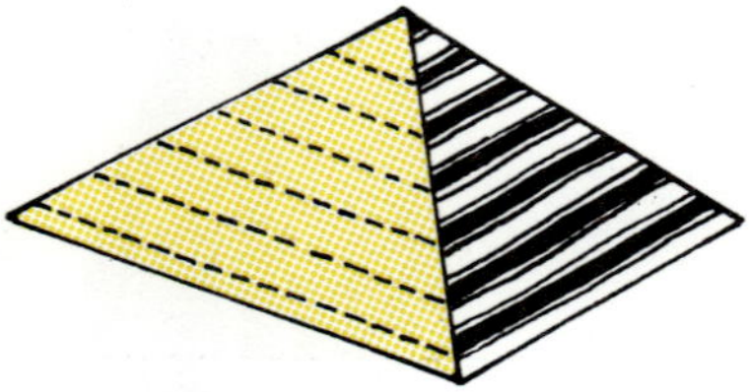

In each case make sure that the edges that you are going to seam together are the same length as each other, so that you will not be left with a piece of material on one side and nothing to sew it to on the other! Paper is not suitable for these sewing experiments. Pieces of old material are much better; they will not tear when you turn your models inside out so that the rough edges of the seams are hidden on the inside.

These 'balls' are made of sections, sewn together. What shape is each section? How many sections make each ball?

It's fun to measure friends and plan out a garment that will fit each one. A 'draped' shape that hangs loosely from the shoulders is easiest to start on. Then see whether you can plan shapes that will fit closer to the body.

Of course, you will make some mistakes, but don't worry. The best thing about mistakes is that you don't make them again and though they are depressing they usually teach you something.

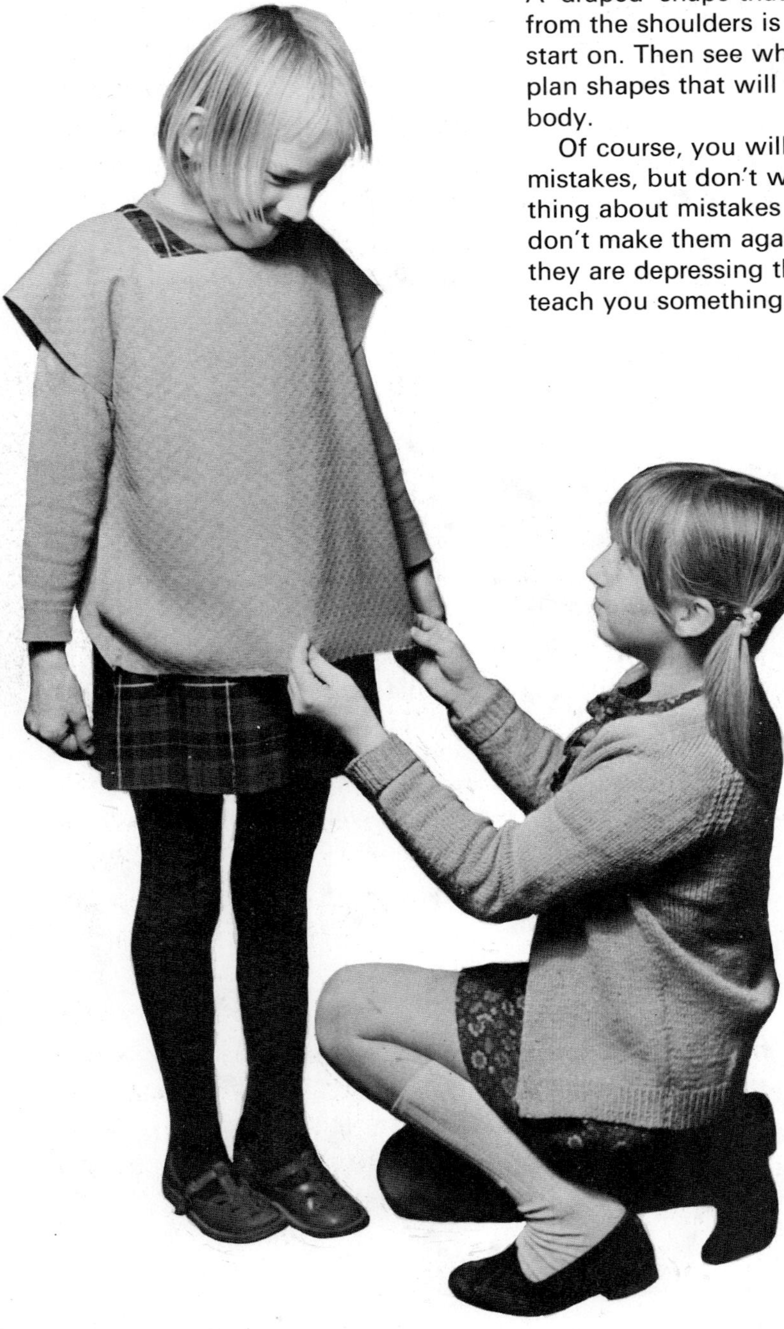

It was a long time before man found a way of making sheets of material to use instead of animal skins. This he did by inventing the spinning of wool (or of flax, perhaps) into thread, and then weaving the thread into cloth. Much later, other kinds of thread came into use, and today we can buy materials made of various kinds of 'natural' thread, such as cotton and silk or 'man-made' fibres like nylon. Most of our clothing is made of thread that has been woven or knitted into flat sheets which are cut into shapes and sewn together to fit the body.

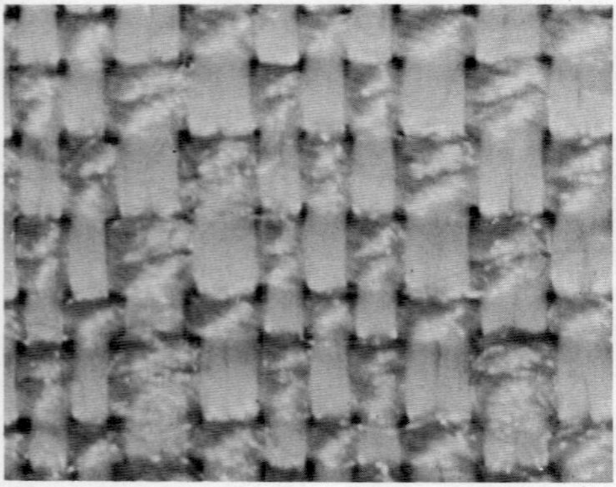
Woven fabric

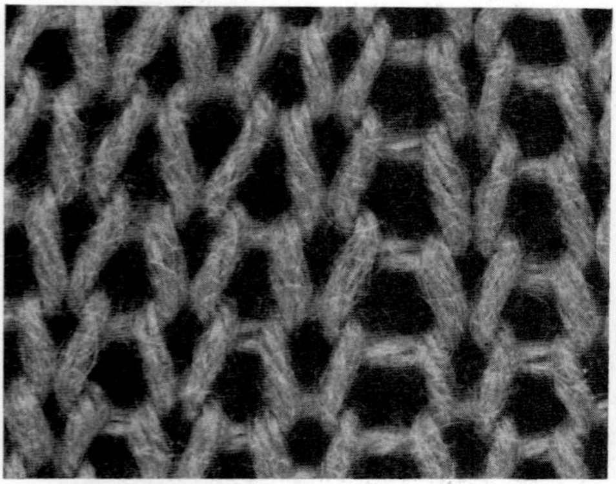
Knitted garment

If you are interested in finding out what *length* of thread is needed to make a garment, I will give you some hints to help you to work out how to do it. When a person is going to *knit* a garment, a printed *pattern* is used. On this pattern is written how much wool will be required. If a square centimetre of *woven* fabric is cut out and the threads are carefully separated and counted, you would know how many one-centimetre lengths go to make *one square centimetre* of the garment. This is best done with coarse material (made of thick thread); otherwise it will be too fiddling a job.

'Ready-made'

It is possible for a tailor or dressmaker to measure you and make a garment specially for your shape and size but this takes quite some time and is expensive. This problem has been solved by what is called 'mass production'. Most of our clothing is made in large factories which have marvellous machines that work many times faster than a human being and do all kinds of tasks, even down to sewing on buttons.

One of the most remarkable of the many machines used is the stock-cutting machine. Many layers of cloth are laid on a special kind of table. The shape that is needed is put on top and then a huge, fast-moving knife slices through all the layers of cloth at the same time, making exactly the same shape as the top pattern. This saves many hours of work.

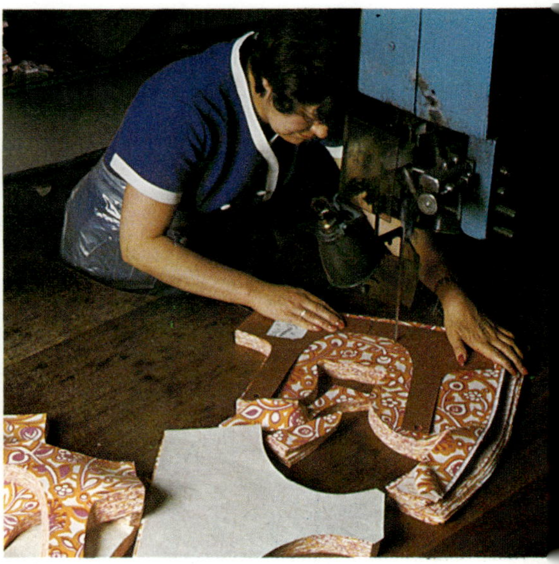

You will remember that in the section on the height of chairs, I talked about the *average seat height* of all the children in the class. I said that this average would be somewhere between the highest and the lowest seat heights of the children who were measured.

If you were to make a garment that was average for *all* people, it would be just the right size for *some* but there would be *many* for whom it would be either too large or too small. Therefore, mass-produced garments are made to the average, not of *all* people, but of *sets* of people who are roughly of the same size.

When you go to buy 'ready-to-wear' clothes, you are sometimes asked what size you are. Most people, for example, know the size shoes they wear. This size is a 'code' number. The man who takes size $8\frac{1}{2}$ and the child who takes size 12 realize that these are not the number of centimetres in the length of their feet! But if a man buys a shirt, the size is the number of centimetres around his neck. Why should it be his neck and not his waist or his chest?

All these garments are the same size, but *people* are not all the same size.

So the factory makes garments in different sizes. But they make a large number of each.

You can buy the same style in different sizes.

People are not as particular about how some clothes fit as others. Pyjamas, for instance, sometimes only come in three or four sizes. Why is it that people are concerned that their suits should fit exactly while their pyjamas are often too big or too small? Recently people have taken more care about *all* their clothes fitting better, even pyjamas. But if you had measured some of the old-fashioned loose clothes you might have found that they were such a rough fit that they went up in *proportion.* You might just find this if you measure the length and breadth of the 'draped' shape clothes you made on page 32 for different sized friends. If things are in *proportion* it means they have the same shape even though the size changes. The picture of the rectangles will help to show how.

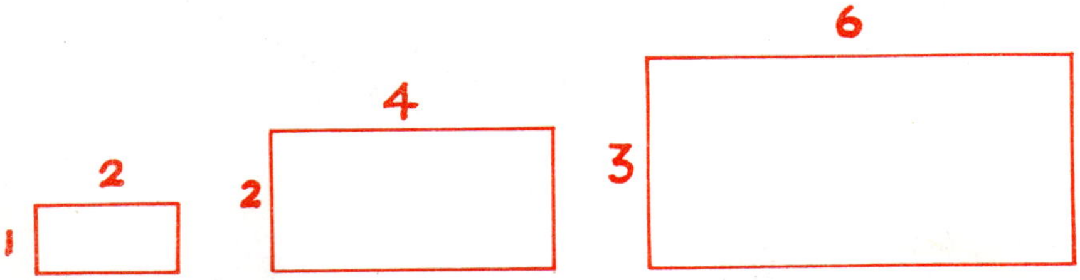

Apart from being rectangles, what else is the same or *similar* about them? As you have probably noticed, in each case the long side is *twice* as long as the short side (or, the other way round, the short side is *half* the length of the long side).

If we were to take any one of them and increase the short side by any length, as long as we increased the other side by *twice* that amount, the shape of the figure would be the same although the *size* had grown larger. Just to check, take the middle rectangle and increase its short side by 7. Now it is 9 long. (It could be 9 anything – centimetres, feet; it does not matter.) But if you have increased the short side by 7, you must increase the long side by *twice* 7 (that is 14). Draw it if you are not sure and check that the shape is still the same. These are called rectangles in the *ratio* of 1 to 2. You should try the same with rectangles in different ratios such as 1 to 3 and 1 to 4 (remember that for every one unit the short side grows, the long side must grow either 3 or 4 units). Then try with the rectangle 2:3. Here the long side is $1\frac{1}{2}$ *times* the length of the short one. Ask for help if you need it.

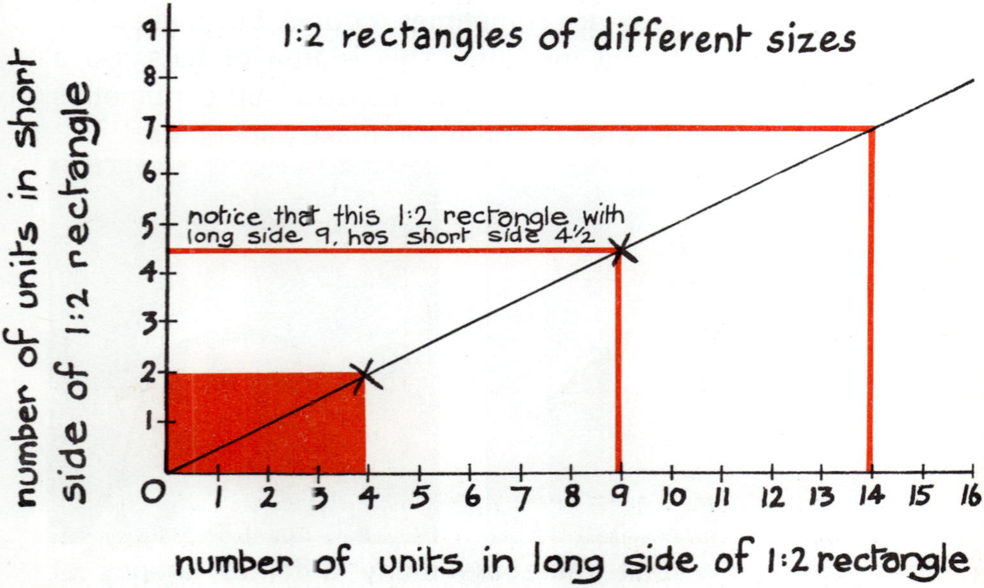

But here is a graph that shows the growth in size of a 1:2 rectangle. Notice that the *intersections* (where the crosses are) go up in a straight line. If you put in more intersections for any other rectangles which have one side twice as long as the other, they will keep on this line. This means that the *shapes* are the same although their scale, or size, may be different.

Now can you make a similar graph for some of your clothes, or anything else you think looks in *proportion.* Instead of your *axes* being called 'long side' and 'short side' they will be 'length' and breadth. You will probably need help to see how to fit the numbers of centimetres into your sheet of graph paper. What this experiment asks is: Do the objects you have measured keep the same shape although the sizes are different? The answer will depend on whether the three intersections are in a straight line.

'If the cap fits'

We said that sizes are sometimes 'codes', but sometimes they are the actual measure. This is true of hats. So a person who wears a size 50 hat measures 50 centimetres round the head. Most children can make paper hats.

We know that there are many different shapes of hat. In fact there is practically no shape, however strange it may look, that has not been used as a hat somewhere or other. But here are some shapes that are quite familiar:

Odd hats

All these hats seem to have circular bases but, in fact, circles would not fit our heads comfortably. You will see this if you get up *above* someone's head and look down on it. It will look like this:

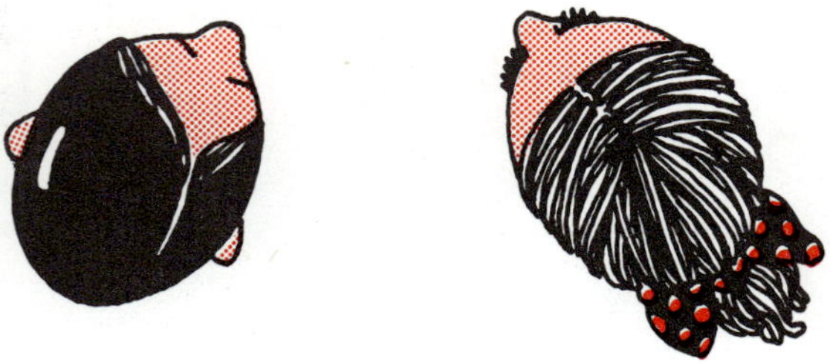

It is fun to make hats out of paper or card and it is very interesting to plan the shape so that it is a good fit. I will suggest how to make a few basic shapes and you can make up others for yourself.

Bend a strip of strong paper or thin card (not more than 2 centimetres wide) round the head, overlap about 2 centimetres and stick. Next, stick a similar strip from front to back. Then the same from side to side (also stick where the strips cross).

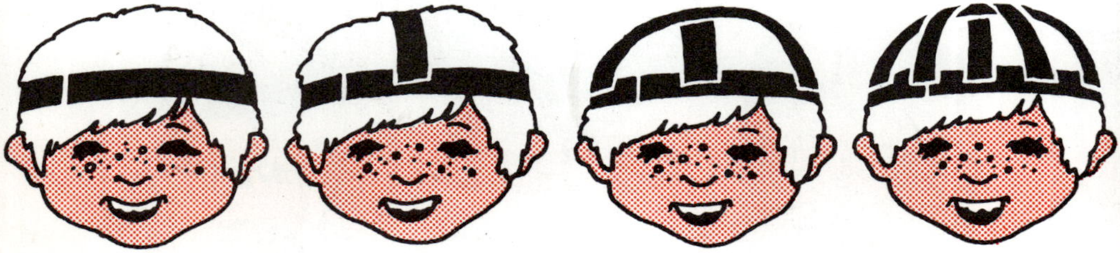

Now fill in the four open spaces with short strips. You will have made a sort of 'skull cap'. This can now be added to and made into a peaked cap by cutting out a peak shape from stiff cardboard and sticking it into position. You will need to experiment with different shaped peaks until you find the one you want. Notice the shape of the tabs. These 'pointed teeth' are necessary whenever you are joining two curves.

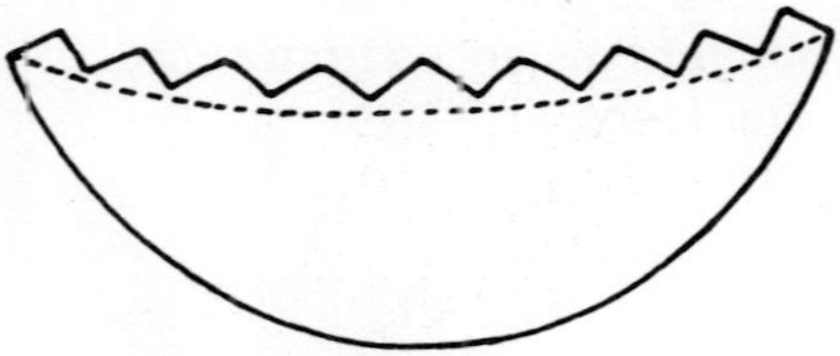

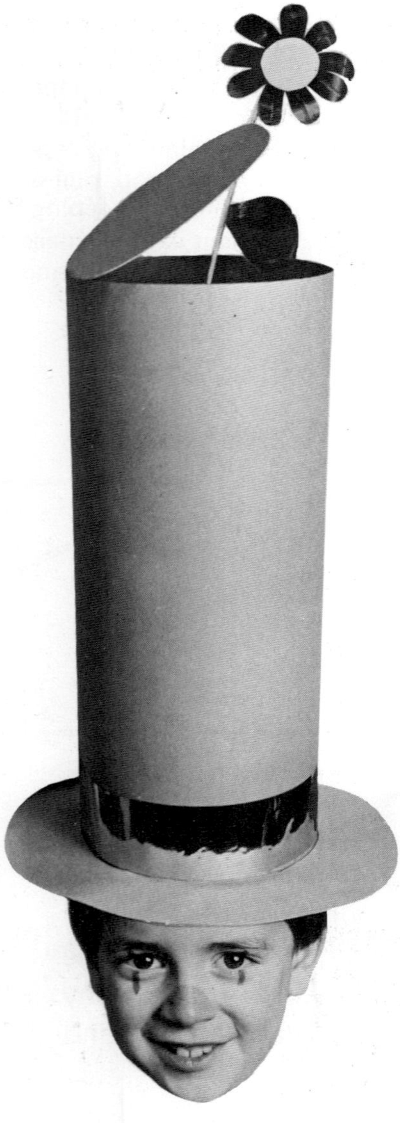

A 'cylinder' hat has a stiff brim going all the way round, so you need three shapes:

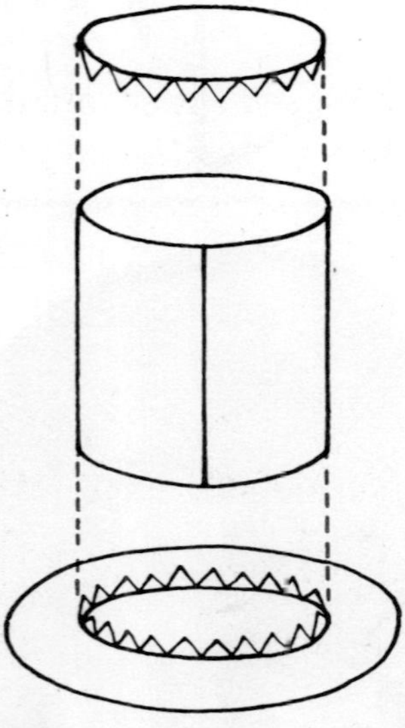

It will be best to make the brim first. Do you see that the shape you cut out of the middle can be used for the 'lid' at the top? This shape is called an *ellipse* and if you can find a way of drawing one neatly to the size of the head, your hat will look (and fit) better than if you sketch it.

Drawing an ellipse

A sheet of drawing paper is pinned to a flat board. Two nails are hammered in a little way and a loose loop of strong thread or thin string is placed round them. You put your pencil in the loop, pull outwards until it is taut, then draw a line right round, keeping the thread taut all the time. Experiment with the nails at different distances from each other and with larger and smaller loops.

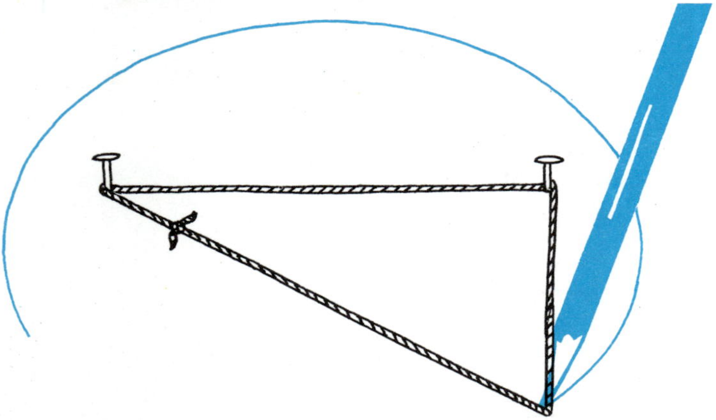

There are many other kinds of hat. With a little experimenting you will be surprised how many you can make, and make to *fit*.

You in the streets and parks

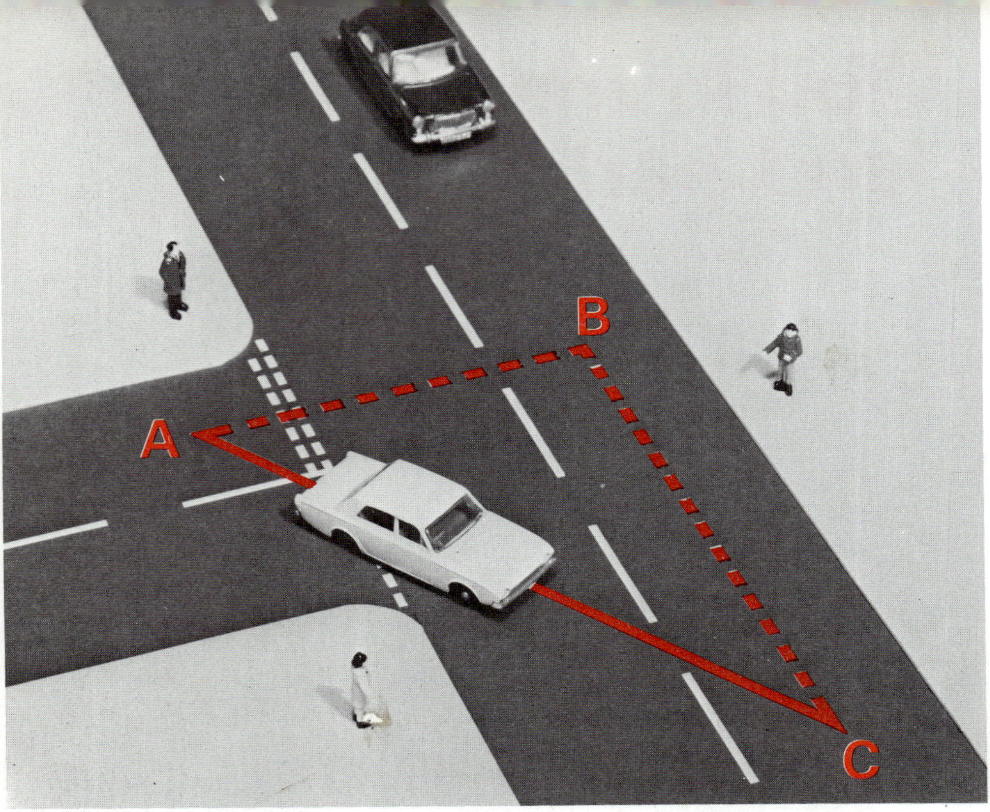

It's becoming more and more difficult to travel on the roads safely. Mathematics is one of the ways which can help us to understand the dangers of road travel.

People often want to get somewhere more quickly by going fast or taking a 'short cut'. This motorist, for example, is 'cutting a corner'. He is taking the wrong route from A to C and should have gone the way shown by the dotted line.

His *mathematics* was all right. Do you see that the lines joining A, B, and C make a triangle? If he had gone correctly from A to B to C, he would have travelled along *two* sides of the triangle. By going direct from A to C he has travelled along the third side only. Is it true that any two sides of a triangle are longer than the third?

Would you think that it is true of *every* triangle? Try drawing a few triangles and see. (If you can't do that ask someone, or look it up.)

Yes, his *geometry* was all right, and he might save himself a second or two of time. But his *common sense* should tell him that the way he has taken has kept him going dangerously against oncoming traffic for quite a long distance.

Getting across a road safely is far more important than getting further down the road. So that, in this diagram, if you are at A, your route to get to your friend at C is A to B to C, however clear the road may look at the time. Modern vehicles travel very fast and many drivers are tempted to speed down a clear road

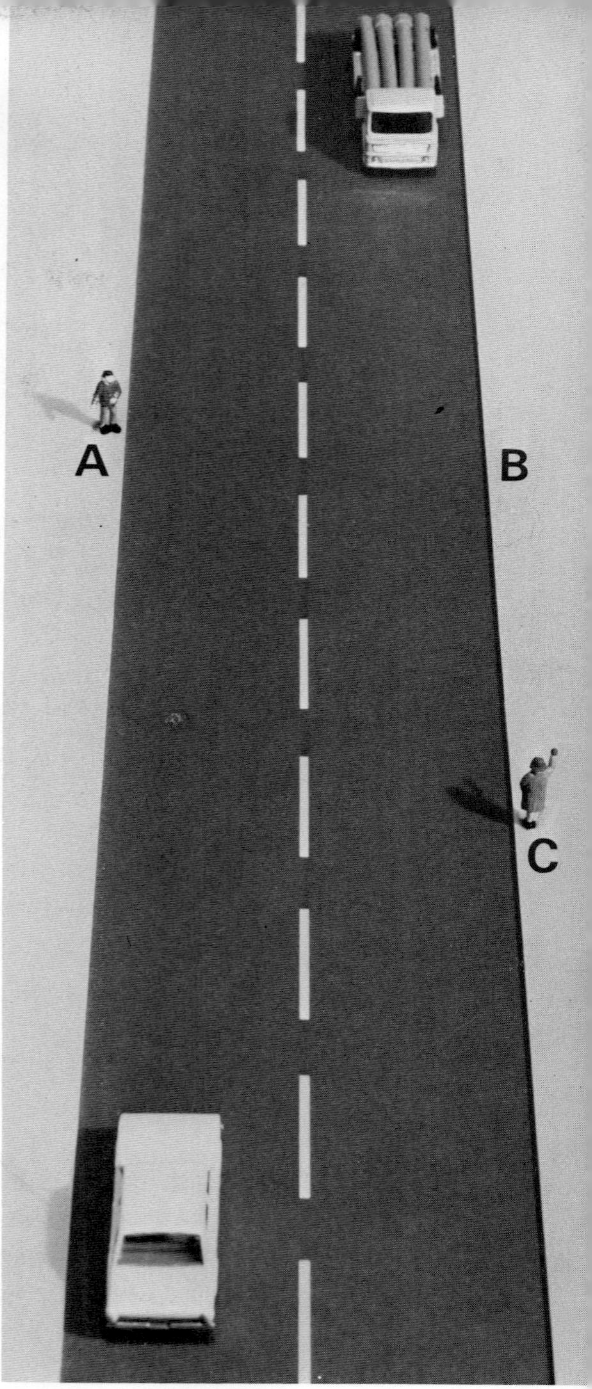

In the Highway Code, drivers are warned that the faster they are travelling, the longer is the braking distance if they want to stop. Is this likely to be as true of human beings as of vehicles? At times children, without thinking, run down a drive or path towards a road. When they have worked up a great speed, they find out, sometimes too late, that they cannot stop themselves from running into the road.

Here is an experiment you may like to try. In an open space, which has no traffic, such as the school playground, chalk a path leading to a 'road'.

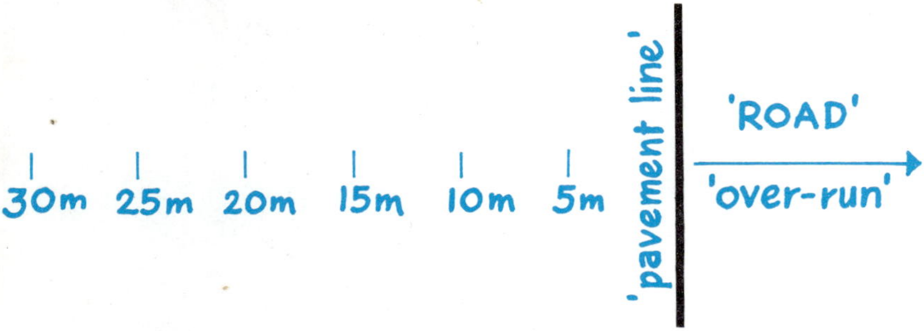

Mark off 'starting lines' as shown. Let one child start at the 5 metres line and run as fast as possible towards the 'road'. As soon as he reaches the 'pavement line', someone must call out 'Stop!' Measure his 'over-run' in centimetres and enter it in a table you have prepared, like the one shown here. Then let him start from the

10 metre line and do the same as before. Continue like this until you have all the distances and 'over-runs' written down. You must make sure that your runner does not begin to slow down as he approaches the 'pavement line' – he must keep up his full speed until he hears the shout to stop.

running distance in metres	'over-run' in centimetres
5	
10	
15	
20	
25	
30	

If you make a graph to compare the length of the run with the length of the 'over-run' into the 'road', then you have made a 'stopping distance chart' and you can compare the shape of it with the graph shown here of the shortest stopping distances of an average car.

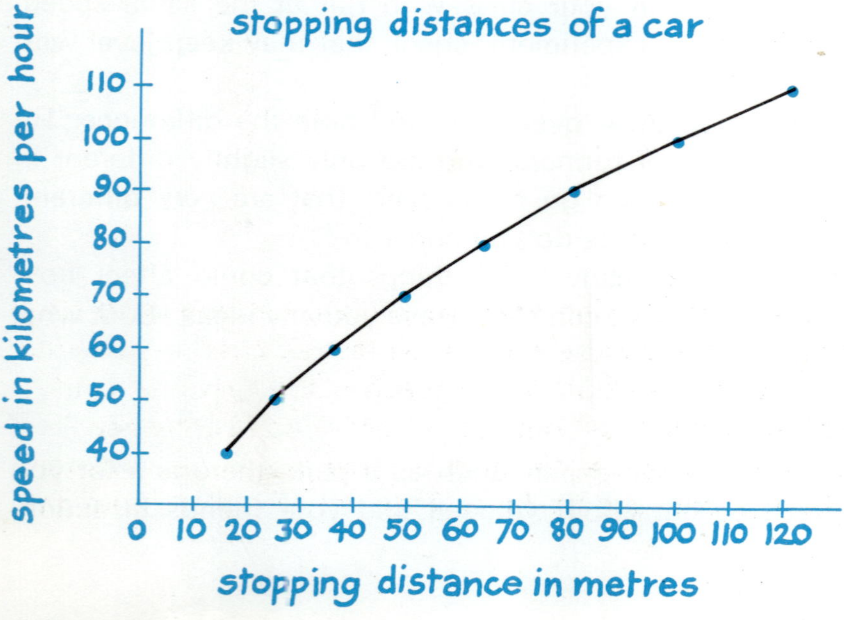

The Highway Code graph shows that a car has an increasing difficulty in stopping as it travels faster. Does your graph line curve the same sort of way?

I hope you have noticed the *differences* in the two diagrams shown here. To begin with, a child is not a motor car! (But there are *some* things in which they *are* similar which you can probably think of.) Then, the *horizontal* axis in one diagram has written beside it 'Speed in kilometres per hour' while the other has 'Running distance in metres'. So, if the shape of your graph is not similar to the other, you may be able to puzzle out why. But if it *is* similar, with the line rising increasingly, then it is probably because the longer run-up you have, the faster you travel, and the more difficult it is to stop.

Is it more difficult to stop if you are a *heavy* person? You can try this out by doing a similar experiment to the last one. You do the same as before but, instead of one person at a time making the run, you have two running together, one heavy person and one light.

They must keep together; so begin with a race to find two people in your class who run at the same speed. In the actual experiment, check that they keep level with each other.

Measure their 'over-runs' and note the difference. Try with a pair of runners who are only slightly different in weight and then go on to pairs that are *very* different. What conclusions do you come to?

There are many other things that could affect how quickly a runner can stop. Have you any ideas about what they might be?

Time trials

In a safe open space, such as a park, there is a lot you can find out about yourself and your friends in action.

With a stop watch, timing clock or even an ordinary watch with a second hand, you can compare, for example, your sprinting speed over a short distance and work out what that speed is in kilometres per hour. This means the top speed you can travel on your feet and does not include the 'running up' to get to that speed nor the slowing down when you are getting tired.

You can probably keep up your best speed for a quarter of a minute and here is a way of finding out how many metres you can run in that time.

Mark a starting place and make sure you have a clear stretch for your run. Take a trial run of 10 seconds just to find out roughly where you are going to end up. That is where your friend with the watch should stand. Another friend should be at the starting-point with some kind of flag – a handkerchief will do.

You do not start running at the starting-point – you want to get up your top speed before the timing begins – so move back for whatever distance you think you need. Now the starter holds up his flag. This is a sign to the timer to get ready. Then off you go! As soon as you pass the starter, down goes his flag. Exactly as he sees the flag go down the timer begins the 10-second count. At 10 seconds he calls out 'Stop!' and you can slow down to finish, but a *third* friend marks where you were at the time of the shout to stop.

You may have to do this several times to get the arrangement right but, when it is done, you will have a marked distance for your sprint. This is best measured with a metre trundle wheel but you can, of course, use some other method.

Now you work out your speed in kilometres per hour. You ran for 10 seconds ($\frac{1}{6}$ minute) and covered — metres. So for 1 minute at the same speed you would run — metres. In 60 minutes (one hour) — metres.

This last statement shows your sprint speed in metres per hour.

As a thousand metres make one kilometre: Your sprint speed is — kilometres per hour.

If you find all this a bit complicated when you try to work it out, ask your teacher to help.

Up we go!

Nowadays, where we have to go from one level to another, for example, at sports stadiums, on road bridges and in subways, there are two methods in use: there are the usual steps but there are also *ramps,* or long slopes, so that prams, wheeled shopping baskets and invalid chairs can be moved about more easily.

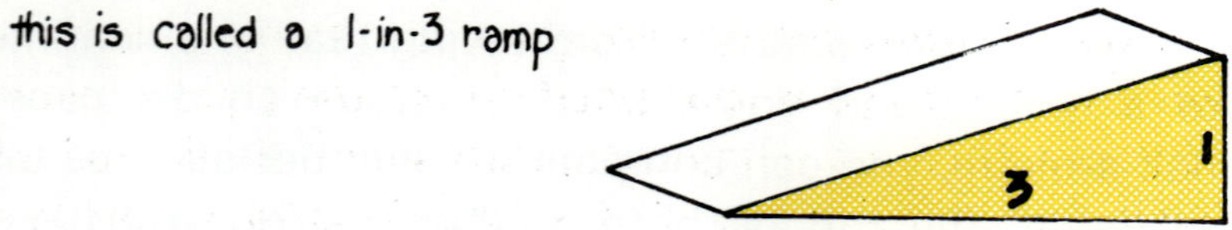

Do you think that even the 1-in-6 ramp is gentle enough for prams and invalid chairs?

If you make a drawing of a flight of stairs and a ramp, you will find that, to be sensible, you will have to make the *angles* at which they go up very different. In the drawing shown here, they rise at the same angle and you can see how impossible it would be to push a pram up the ramp! But if you make the ramp at a shallower angle, you lengthen the distance that has to be walked. The whole idea of the ramp is to make things *easier.* Perhaps if you draw a steps and ramp diagram, but give *your* ramp a *gentle* slope, you can measure the distance along the ramp and compare it with the distance walked up the stairs.

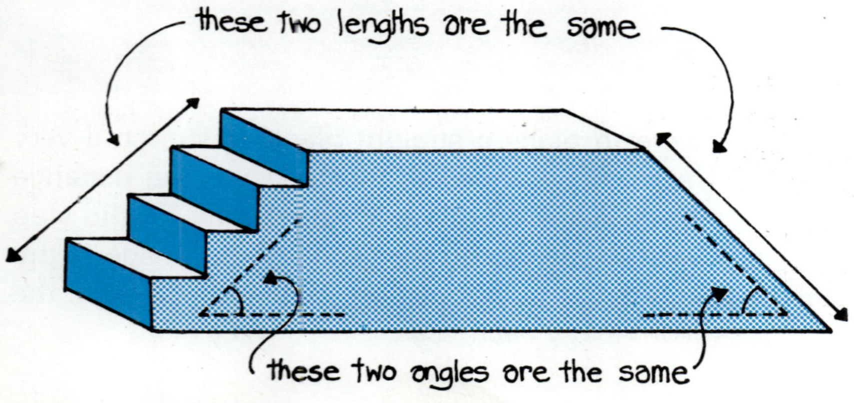

If you make yours a scale drawing (say one centimetre to a metre) and you take an average child's pace as half a metre, you can compare the number of steps taken to get to the top by both methods. (You will have to measure along the ramp and work out how many half-metre lengths it contains, but each stair will be one step for the child.)

Talking about steps and stairs, have you noticed the worn steps of very old buildings? They have been trodden on so many times over the years that gradually they have changed their shape. Look at the picture. Can you see which part of the steps has been most trodden on.

If you were to place a straight plank or stretch a very taut cord across a worn step and measure the distance down from the plank, you would know how far the step has been worn away. If you then found out the age of the building, you could work out how much, on average, the step was worn away each year.

You and sport

We have already said something earlier in this book about running. This is a sport in which you use your own body and nothing else. But there are many sports where we use things made especially for a purpose. Often they are for hitting with. The thing that is hit is usually a ball of some kind.

You will find that the rules for many ball games are very precise about the weight and size of the ball. By experimenting with these different balls, you will find some interesting differences in the speed and distance they can be thrown or hit. If you made a golf ball the same size as a cricket ball, would it be heavier, lighter or the same weight as a cricket ball? Here is one way to find out. First, carefully weigh and note down the weight of a cricket ball and a golf ball. Next we have to find the *volume* of each of the balls. The best measuring unit for this would be the *millilitre* (one millilitre of water, if shaped into a cube, would be just this size). How many millilitres would you expect there to be in the cubic metre on page 18? So we will need a millilitre measuring jar. Prepare a tin or similar container full to the very brim with water and sitting in a dry bowl.

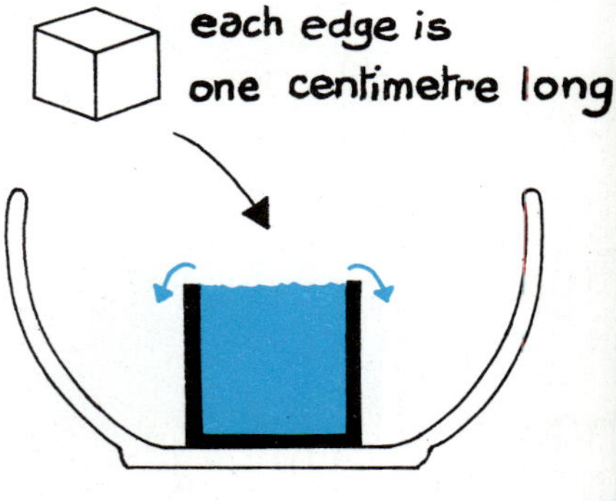

Place one of the balls in the water and make sure it goes down below the surface. Where the ball is now there was water before. Where does this water go? As you put the ball in the water, it spilled over into the bowl. So the water in the bowl is exactly the same *volume* as the ball. Pour this water carefully into the measuring jar and write down the kind of ball and its volume in

Weight in water

We play games in water also, such as water polo. Do you think you are the same weight in the water as you are on land? Look at these two photographs. In one the child is really *in* the water. In the other he is *partly* in the water but partly in the air.
The person to be weighed stands in water reaching up to his chin. He holds the string with his hands at chest level. Keeping them like this he bends his knees and crouches until his head is well below the level of the water. Then the person holding the spring balance gently lifts until the crouched body is lifted slightly off the bottom. The weight of a body in water is shown on the spring balance only while the whole body, including the head, remains under the surface. This makes you think about the body's weight in relation to things like iron or wood. Is it heavier or lighter? And, of course, is it heavier than *water*? Can you find out? You could compare it with your friend's weight in air which you found out earlier.

(Body experiments in water should not be done without your teacher.)

millilitres. Do exactly the same with the other ball. Now make this graph, which shows the weight of a golf ball together with its size. My lines at 6 grammes and 60 millilitres, which meet at B, are *not correct,* because these are not the actual weight and volume of a golf ball. Anyway you have already found out for yourself and written down the *true* weight and volume, so your two lines will be different from mine and your point B will be in a different place on the graph. From your *true* point B, rule a straight line to A. Then extend the other end as far as your paper will let you go. That is your point C. Your A-B-C line makes it possible for you to find what a golf ball would weigh if it were made a larger (or smaller) size. Suppose you wanted to know the weight of a golf ball made to the size of 25 millilitres. You simply draw a line from the 25 millilitre point on the volume axis, straight across until it reaches the A-B-C line. Then draw a line straight down. Where it comes to, will show the weight of a 25 millilitre golf ball. You know the volume of your cricket ball in millilitres. Find out by your graph what a golf ball this size would weigh. What difference do you think such a large golf ball would make to the game of golf?

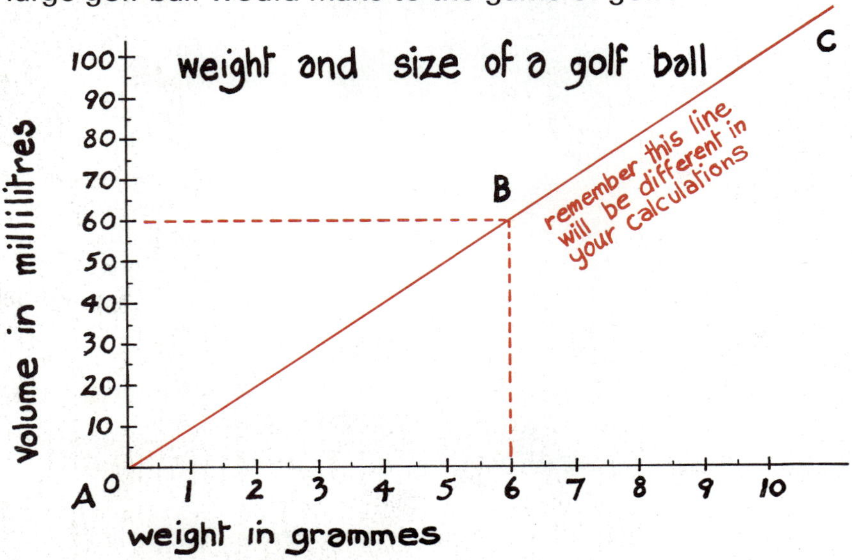

You could try experiments with all kinds of balls. You would soon discover very good reasons why the ball you have to catch or throw is very different in size or weight from the one you simply kick or hit.

Why is a cricket bat used? One hand, even if a specially strong glove were to be worn, is not very strong, and it takes a lot of power to hit a ball a long way. If you want to use the strength of *both* arms together to hit the ball hard, is a cricket bat the best thing to use? Why is a cricket bat the length it is? What would be wrong if it were longer or shorter?

With a cricket bat it is possible to use almost the full strength of both arms. It is just as though your two arms are joined together at the handle like a letter Y and the leg of the Y is as long as it can be, but not so long as to be awkward when the bat is swung. Therefore, using a cricket bat is the same as joining your two arms together and making them much longer than they really are.

This idea of making your arms longer is found in many sports; sometimes it is one arm, as in tennis, sometimes two, as in golf. But always it gives you more *swing* than your arms by themselves would have. The longer the swing, the greater the force when you hit the ball. But a long swing takes time. If you are standing in one place, when you are making your swing, you can lift your 'bat' well back and get the full length of the swing. This is what you see a golfer doing when he uses a heavy 'driver' club. On the other hand, when you are playing tennis, you have to run fast from one place to another to meet the ball. For this and other reasons, a tennis racket must be light and springy. Can you work out any of the other reasons?

By now you can probably think up for yourself experiments to test some of the things we have been thinking about, except for the last one which can be a bit dangerous. A simple but very interesting one is to make a swing-testing pendulum. Here is one way it might be done.

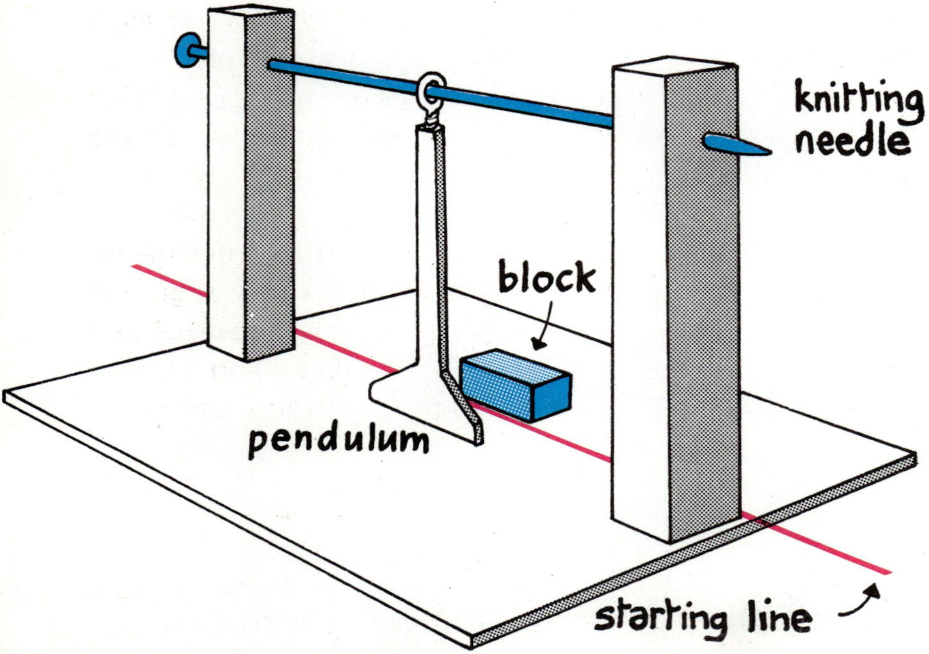

On the starting-line is placed a block weighing 100 grammes or so. The pendulum is pulled up to an angle of about 45° (which is half a right angle) and then let go. It knocks the block forward and you measure how far it has been moved. Next, you put the block back on the starting-line, pull up the pendulum to a full right angle (90°), let go, and measure the distance the block has been moved this time. Next time 135° and then 180°.

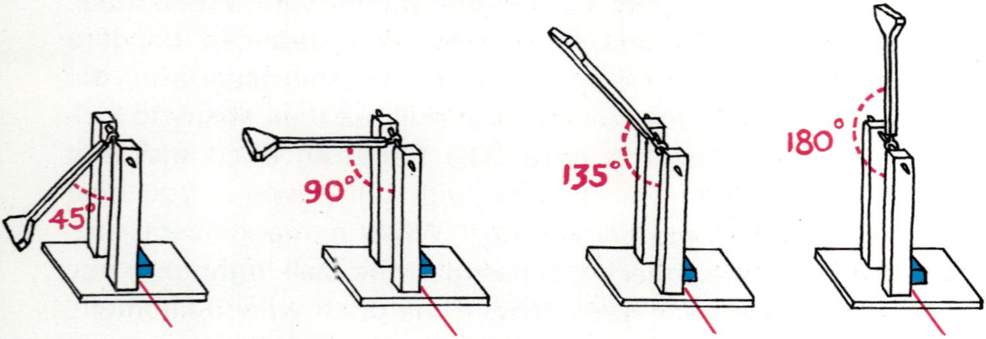

No special measurements are given for this apparatus and it does not have to look exactly like this. If it does not work well enough, plan alterations for yourself. At each increasing swing, measure how far the block travels and see if the increase is about the same each time or whether it varies, getting regularly bigger or smaller. A simple graph would show this:

You will have to decide for yourself how to arrange the numbers on the horizontal axis because it will depend on how hard your pendulum strikes.

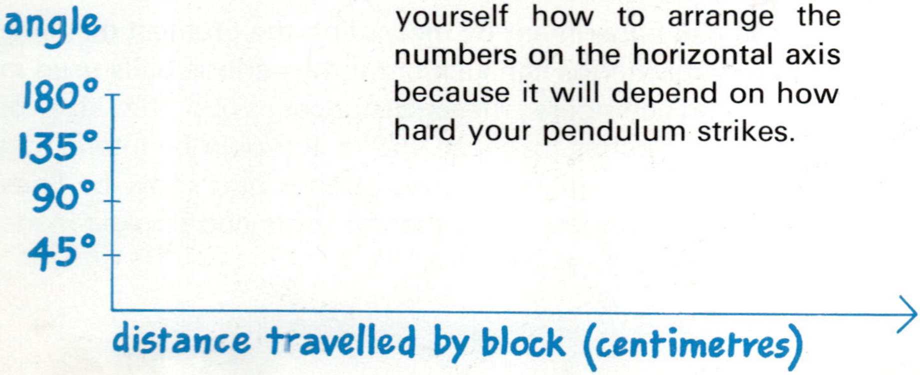

This is a good example of how one experiment leads to another for, when you have found out the answer to the above experiment, the next question that might well occur to you is, 'Will the block travel further if the striker is heavier?' You could double the weight of the striker and find out.

The field of play

The sizes of pitches for different games vary a great deal, but they all depend on what our bodies can do. If a football pitch were half or a quarter the usual size, the players could use quite a lot of skill but their ability to kick hard would not be used. The full-sized pitch makes it possible for players to kick with full power – you hear radio commentators saying: 'What a magnificent kick by the goalkeeper! He has put the ball right into the opponent's goal area!' Now if the pitch were a kilometre long, there would be plenty of room for *enormous* kicks, but as no one could possibly kick a ball as far as a kilometre, there would be no point in having a pitch a kilometre long.

But if a modern football pitch is *related* to the strength of the human body in kicking, how is a *cricket* pitch related to our strength in hitting a hard ball with a bat? Also, if a game is being played by boys or girls, who are not as strong as adults, should their pitches be smaller? And are they?

You can experiment by measuring the greatest distance you or your friends can kick or hit the various balls used in sport and compare these distances with the size of regulation pitches for those sports. It would be interesting to draw scale plans of various pitches and show by lines on them the longest experimental shots and strokes made by yourself and your friends.

This does not apply to all games, however. There are some which depend more on skill in *placing* the ball than strength in sending it a long way. This applies to tennis and, even more so, to table tennis. In these games, a more useful experiment might be to play some games with a much higher, and then a much lower, net than usual and find out what effect it has on you. What difference does it make to the rate of scoring? To make a game interesting, it must not be too easy to score, but neither must it be too difficult – think what it would be like if nobody ever scored in most football matches because goals were three feet wide! What if a cricket wicket were one stump wide? Or netball posts 17 metres high?

Football started, hundreds of years ago, as a game played with a pig's bladder. The people below are playing with a ball but even this game was a long time ago. The idea was to get it, by kicking, or carrying, from your own village into the opposing village. The men of the other village, of course, had to prevent this. The game was so rough and so many fights broke out that laws were passed to prevent people playing 'football'. Can you think of other games that possibly developed from these contests between villages? Even today, football sometimes leads to fighting between players or among spectators.

Little ones and big ones

Young boys and girls, because they are not full grown, are not expected to use adult sized bats, balls, rackets and so on. In fact, *adults* do not all use the same sized personal equipment because *they* are not all the same size. Is there any game or race in which children are expected to perform on the same sized pitch or along the same distance as grown-ups?

If there were a netball match between girls of ten and adults, would it be fair to lower the net, rather than to ask young girls to score through a high net? Look at these two drawings.

The girl is 1·3 metres tall and the woman 1·7 metres tall. How much higher should an adult's netball goal be than a child's? In this case the difference in the height of the two people is 0·4 metres but, of course, all girls are not 1·3 metres and all ladies 1·7 metres. This is my *guess* at the average height of a player in a ten-year-old girls' team and an average-sized woman in an adult team. Do you remember how to find an average?

Netball posts have three possible net heights with one foot between each and the next, so it seems as though the average *tall* player should be two feet taller than the average junior player (say a ten-year-old child).

Does it depend on their heights from the feet to the top of the head, or should it be from the feet to the end of the upstretched arms, because that is the position a netball 'shooter' takes up when trying to score?

If you line up a number of people whose heights increase by five centimetres each time, and draw a line across the tops of all their heads, it would be a *straight line at an angle to the horizontal.* Like this:

Now if you asked each one to reach up with arms stretched, without standing on tiptoe, and you draw a line across the tips of their fingers, you could call this their 'reach line'.

The question is, will the 'height line' and the 'reach line' go up at the same angle?

Of course, to carry out this experiment you do not really have to line up a number of people. The best way is to draw a graph to scale, after making out a 'table' like this:

head height	reach height
1m 40cm	
1m 50cm	
1m 60 cm	
1m 70 cm	
1m 80 cm	
1m 90 cm	

You do not have to start with 1 m 40 cm and end with 1 m 90 cm. Make up your own mind about the range.

Each of the persons in the table will have to be measured for 'reach' height. If you think that people of the same head height might have different 'reach' heights, then measure several of the same height and find out. If they *are* different, you can work out their average 'reach' height and enter that in your table.

Now you can decide on the scale for your graph. It can be whatever size you like. It could be one centimetre for every 10 centimetres. You might want it smaller, or you might even want it life-size. If you draw or paint the line of people, it can make an amusing picture-chart, but be careful to get the lengths right and make your 'head height' line and 'reach height' line distinct and the horizontals *parallel* with the 'floor' on which the figures are standing. And when you write your account of an experiment you have to bear in mind the question you asked to begin with. This time it was 'Do height lines and reach lines go up at the same angle?'